Chalk Up Another One

Chalk Up Another One

The Best of Sidney Harris

Foreword by Leon M. Lederman

Rutgers University Press
New Brunswick, New Jersey

All the cartoons in this book have been previously published and copyrighted by the following periodicals: *American Scientist, Management Review, Medical Economics, Medical Tribune, Medical World News, Private Practice* and *The Wall Street Journal.*

They have also appeared in the following books: "What's So Funny About Science?", "Chicken Soup and Other Medical Matters", "All Ends Up", and "What's So Funny About Computers?", all published by William Kaufmann, Inc., ©1977, 1979, 1980, and 1982, and "Science Goes to the Dogs", published by ISI Press, ©1985.

Library of Congress Cataloging-in-Publication Data
Harris, Sidney.
    Chalk up another one : the best of Sidney Harris / foreword by Leon M. Lederman
          p.   cm.
      ISBN 0-8135-2260-9
      1. Science — Caricatures and cartoons.   2. American wit and humor.   Pictorial.   I. Title
NC1429. H33318A4   1992
741.5'973 — dc20   92-23430
                                                                  CIP

Printed in the United States of America

For Bill Kaufmann, who first published
96% of these drawings in book form.

Foreword

There is an honorable tradition that relates humor and science. One doesn't have to go back further than to Democritos of Abdera, the inventor of atoms, also known as the Laughing Philosopher, who was moved to mirth by the follies of mankind. Curiosity about the world and humor enter the evolutionary stream for reasons that are not at all obvious. One theory, illustrated by the original film "MASH," asserts that so much of human behavior borders on the wildly irrational that the only way to preserve sanity is by contrived insanity — i.e., humor. Archimedes ran down the street naked shouting "Eureka" in his excitement at discovering the law of displacement but we don't know if it was to be funny or whether he was trying to market his discovery.

Sidney Harris has been there to rescue us all from the insanity of the daily headlines starting with his first science collection, "What's So Funny About Science?" to this best-of-the-best volume, "Chalk Up Another One." Harris has become the Laureate of Laughs, the acknowledged humorist of choice among scientists, and the public that laughs at them. As you turn the pages of this collection you will begin with a giggle (the couple in search of an obscure and romantic restaurant called Fibonacci's) and proceed to helpless, semi-hysterical laughter at the Einsteinian dilemma of whether E should be equal to ma^2, mb^2 or mc^2. We groan over the injunction that because of budget cuts we must achieve our breakthroughs early in the week, and maybe even long for the good old days when the periodic table consisted of earth, air, fire and water, and when we didn't compete for funding with 90 percent of all the scientists who ever existed.

Another aspect of the deep connections between science and humor which underlie Harris' commentary has to do with the phenomenon (also touched on by Lewis Thomas in "Lives of a Cell") that when we *finally* do understand something, some puzzle that has baffled generations, when we get to that Eureka moment, we tend to be engulfed in laughter at the simplicity of it all — "Of course it is this way! How could it be any other way?" The humor is part relief, part self-mockery (How could we be so dumb?), and part just the exuberant joy of discovery.

In a sense, each cartoon is a scientific statement. Like all our publications, some are more significant than others. In each, a facet of our lives as scientists stands revealed — absurd, ludicrous, but each with its atom of truth. For scientists between classes or during a quiet night shift on the accelerator, there is nothing better to give you proper perspective on your science and your fellow scientists. For observers of the scientific scene — you "Brief History of Time" graduates — the key to our conquest of matter, energy, space and time is provided by this profoundly hysterical collection of cartoons.

Leon M. *Lederman*

August 1992

"I think you should be more explicit here in step two."

INCONCLUSIVE EXPERIMENT:
PAVLOV'S CAT

"You have a choice of three courses. You could increase speed somewhat and retain your comprehension, you could increase speed considerably and reduce comprehension, or you could increase speed tremendously and eliminate comprehension completely."

"As I understand it, *they're* in danger of becoming extinct, too."

"Actually, they all look alike to me."

"Matthew want eat. Get banana. Get bread. Get milk."

"Just between you and me, where does it get enriched?"

"This is the part I always hate."

"It certainly becomes uncomfortable when the pollutants
are up to 990,000 parts per million."

"Why, it must be somatotropin, the growth hormone!"

"Well, Gottfried, news from the cloning front, I see!"

"Don't bother Daddy now. He's singing."

STALACTITES
GROW FROM CEILING

STALAGMITES
GROW FROM FLOOR

PLEASE **DO NOT**
ASK GUIDES
WHICH IS WHICH

"What I'd really like to do, of course, is just find a cure for the common plague."

"Life, yes — but as for intelligent life, I have my doubts."

"But we just don't have the technology to carry it out."

"I can't *stand* this waiting. Couldn't we be six-year locusts?"

"Bunsen, I must tell you how excellent your study of chemical spectroscopy is, as is your pioneer work in photochemistry — but what really impresses me is that cute little burner you've come up with."

"What's wrong with white mice and guinea pigs?"

"Oh, for Pete's sake, let's just get some ozone and send it back up there!"

"I tend to agree with you — especially since $6 \cdot 10^{-9}\sqrt{t_c}$ is my lucky number."

"How do you want it — the crystal mumbo-jumbo or statistical probability?"

"There's *some* light coming from it. We'll just have to assume it's a dark gray hole."

"But technology has created an information explosion —
everyone *does* have to talk at once."

"IT JUST ISN'T WORKING. WHAT SHALL WE DO?"

"Stick to it. There's a future in cryogenics."

"What's the big deal about solar heat?"

"Just don't think about it. We've always been carnivorous, and we always will be carnivorous."

"But Gershon, you can't call it Gershon's equation
if everyone has known it for ages."

"I wouldn't worry. With continental drift, Africa or South America
should come by eventually."

"What it is is a giant kidney."

"There's a 60% chance of 20% acid-rain and
a 40% chance of 30% acid-rain."

"Keep in mind that, like everyone else, I use only ten percent of my brain."

"Although humans make sounds with their mouths and occasionally look at each other, there is no solid evidence that they actually communicate with each other."

"Oh, oh — looks like a blue shift."

"I think we should ask Zimmer to do those experiments.
He's a Capricorn."

"Remember when there was all that fuss about recombinant DNA?"

"Apparently, Mr. Fradkin, evolution is a two-way street."

"Garfield, I think I know why we've been receiving so few commissions."

LEFT BRAIN
DOMINANCE

RIGHT BRAIN
DOMINANCE

S Harris

"Take it from me, and come back. The future is definitely on land."

"Now that we've got *this* wrapped up, I'd like to get into math."

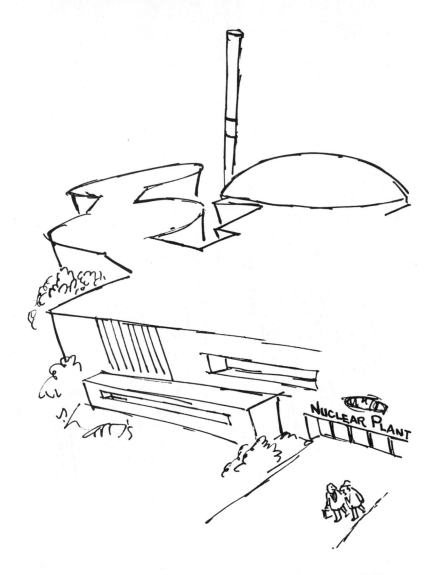

"Of course it's perfectly safe. Any accident would be in complete
violation of the guidelines established by the Federal Nuclear
Regulatory Commission."

"The beauty of this system is that there are a few small errors programmed into it, which helps to avoid total depersonalizations."

"The big bang? Believe me, it was very, very, very, very, very, *very* big."

"He's the typical American mouse — likes a drink before dinner, smokes a little, watches TV..."

"Sure, we're dealing with tiny particles but your formula
is just a *symbolic* representation."

"Our problem, once solar energy is in operation, is to find a way to have
the citizens whose homes are heated by the sun
continue to pay *us* every month."

"There's another hereditary disease that runs in the royal family.
Your grandfather was a stubborn fool, your father was a
stubborn fool, and *you* are a stubborn fool."

"He was working on a theory of entropy, and developed
a severe case of it himself."

"If we ever intend to take over the world, one thing we'll *have* to do is synchronize our biological clocks."

"The periodic table."

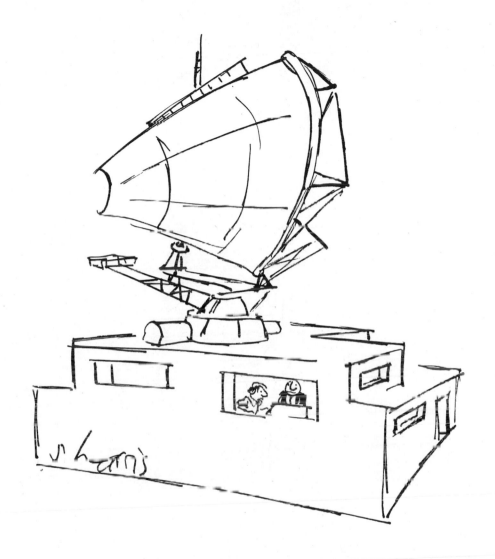

"As I read it, we're receiving a message from outer space telling us to
stop bombarding them with unintelligible messages."

"You both have something in common. Dr. Davis has discovered a particle which nobody has seen, and Prof. Higbe has discovered a galaxy which nobody has seen."

"That's *it*? That's peer review?"

"Frankly, I don't see how we can keep it burning through eternity."

"Looks like a virus."

"I love hearing that lonesome wail of the train whistle as the
magnitude of the frequency of the wave
changes due to the Doppler effect."

PROBABILITY

If you have 5 dogs, 3 will be asleep.

"This is not what we meant, Snider, when we asked for
a thorough study of the laws of gravity."

"The last I heard, Medwick was working on a
model black hole in his lab."

"What's most depressing is the realization that everything we believe will be disproved in a few years."

"I thought continental drift was much slower."

"He claims to be a specialist, but I think he just has a one-track mind."

"I've been trying to trace my roots, but after a couple of generations, it goes off into a different species."

"*I* won? I didn't even know there *was* a Nobel booby prize."

"It's the latest miracle fabric: 40% dacron, 40% orlon, 20% recombinant DNA."

" 'Look', I would say to Leonardo. 'See how far our technology has taken us.'
Leonardo would answer, 'You must explain to me how everything works.'
At that point, my fantasy ends."

"Whatever happened to *elegant* solutions?"

"I can't complain. Last week they had me on martinis."

"Apparently some of the additives cause a nerve disorder, but some of the other additives cure it."

"My big mistake was going into cosmology just for the money."

Chapter 7. THE STRUCTURE OF THE NUCLEUS OF THE ATOM

"What?" exclaimed Roger, as Karen rolled over on the bed and rested her warm body against his. "I know some nuclei are spherical and some are ellipsoidal, but where did you find out that some fluctuate in between?"

Karen pursed her lips. "They've been observed with a short-wavelength probe..."

"This place is all right. Two more weeks and I'll be a molecular biochemist."

"If you want fiber, Madame, I suggest you eat the menu."

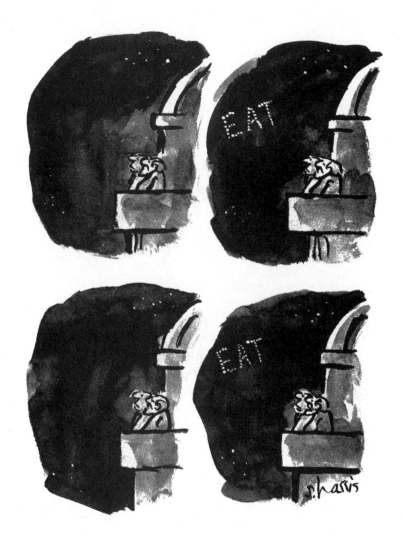

"I don't care what it looks like — they're pulsars."

"We should be thankful. What if oil and water *did* mix!"

"I've tried it. Kicking doesn't work. There must be some *other* way to get oil out of shale."

PROBABILITY LABS

USUALLY OPEN 9-12
OFTEN OPEN 1-5

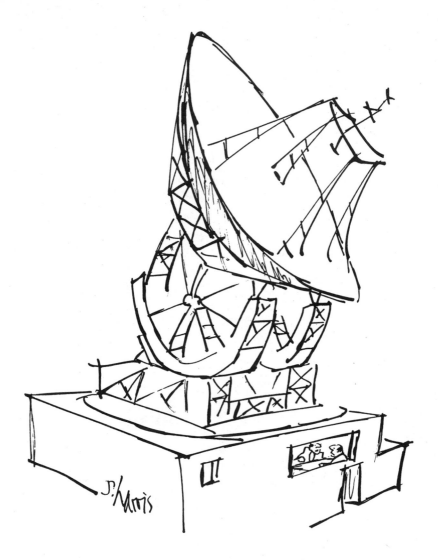

"As I understand it, they want an immediate answer. Only trouble is,
the message was sent out 3 million years ago."

"That wraps it up — the mass of the universe."

"You mean *Casey's* book on Hamlin's Syndrome will be out before *my* book on Hamlin's Syndrome?"

"You'll like this flock. We do the regular migrating twice a year, and then we take lots of these side trips."

"This must be Fibonacci's."

"The forward thrust of the antlers shows a determined personality, yet the small sun indicates a lack of self-confidence..."

"I can remember when all we needed was someone who
could carve and someone who could sew."

"It may very well bring about immortality, but
it will take forever to test it."

"We have reason to believe Bingleman is an irrational number himself."

"What do you expect, since 90% of all the scientists
who ever lived are alive today."

"There was a time I thought humans were as smart as we are."

"Now, if we run our picture of the universe backward several billion years, we get an object resembling Donald Duck. There is obviously a fallacy here."

"Just in case it doesn't work, we'd like you to come up with some uses for ten million gallons of salt water every day."

"But you can't go through life applying Heisenberg's
Uncertainty Principle to *everything*."

"Do your stuff — you're on microscope."

"However, it's excellent pseudoscience."

IDENTICAL TWINS | FRATERNAL TWINS

J. Harris

"I don't know what it measured. The Richter Scale is down there."

"When you're young, it comes naturally, but when you get a little older,
you have to rely on mnemonics."

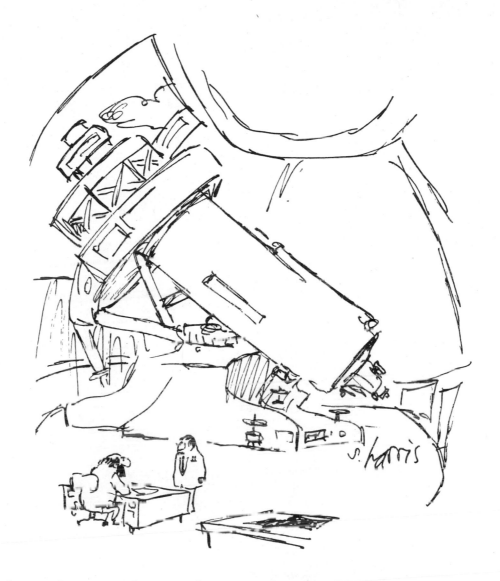

"It's always the same thing — the sun, a few clouds, and that's it.
I'd like a transfer to the night shift."

"My problem has always been an overabundance of alpha waves."

"You know what I hate about this place? The heavy water."

"It sounds like an implosion."

"Son, I hear you failed genetics."

"Perhaps, Dr. Pavlov, he could be taught to seal envelopes."

"I have a feeling it's too soon for fossil fuels around here."

"Another one uninhabited. That's three down
and several hundred billion to go."

"Due to a tightening of the budget, we are forced to curtail our overtime and weekend schedule, and request that all major breakthroughs be achieved as early in the week as possible."

"You don't seem to understand, Prescott. We're not trying to cure diseases occurring *only* in guinea pigs."

"Every once in a while I just like to unwind with a little addition and subtraction."

"The devil with the food chain. I *like* mercury."

"Actually I started out in quantum mechanics, but somewhere along the way I took a wrong turn."

"One advantage of living near a binary star would be
a sun tan in half the time."

"On the other hand, my responsibility to society
makes me want to stop right here."

"Somehow I was hoping genetic engineering would take a different turn."

THE MEANING OF LIFE

S. Harris